AI For The Culture: How Emerging Tech Is Leveling The Playing Field

Nyjal J. Drayton

Table Of Contents

Acknowledgements

This book is a reflection of not just my knowledge and experiences but also the support, encouragement, and wisdom of the people around me.

First and foremost, I want to express my gratitude to my family and close friends who have continuously believed in me as a person. Your support is a big fueler toward my ambition.

A special shoutout to my close friend and mentor, **Vincent Hunt***, Founder, CEO, and Agent 001 at* **The Bureau of Creative Intelligence***. Vincent has been one of my biggest inspirations on this journey. His leadership, innovation, and vision in the tech space continue to push boundaries, and his impact is felt far beyond just those who work alongside him. Thank you for your guidance, wisdom, and for being a true pioneer in this space.*

To my other mentors, colleagues, and those who have inspired me along the way–thank you for sharing your insights and pushing me to think bigger, dream bolder, and execute with precision.

To the communities I serve–this book is for you. My mission has always been to bridge the gap, to provide knowledge that empowers, and to create opportunities where they may not have existed before. I hope this book helps you harness the power of AI to elevate your skills, businesses, and careers.

Finally, to every reader who picks up or listens to this book–thank you. Your curiosity and desire to grow are what make this work meaningful. AI is a tool, but you are the innovator. Keep learning, keep building, and most importantly, keep pushing forward. – With appreciation, **Nyjal J. Drayton**

Preface

AI isn't the future, it's the present. And whether we realize it or not, it's changing the game in ways that go way beyond just tech. It's shifting how we work, how we create, and how we compete in the marketplace. But here's the real question: **Are we taking full advantage of it?**

This book isn't about theory, hype, or tech jargon. It's about **real, practical ways AI can be used to level up your career, business, and creative hustle—right now.** It's for the students figuring out their next move, the entrepreneurs looking for an edge, the 9-to-5 grinders who want to work smarter, not harder. AI is already being used to automate tasks, generate ideas, and streamline entire businesses—so why wouldn't we be using it too?

But let's be real: AI isn't perfect. There are risks, biases, and ethical concerns that come with it, and ignoring those would be just as irresponsible as ignoring AI altogether.

That's why this book doesn't just teach you how to use AI— it also breaks down **how to use it responsibly, strategically, and in a way that actually benefits you.**

At the end of the day, this isn't about AI replacing people. It's about people who know how to use AI **outpacing** those who don't. If you're ready to be on the right side of that equation, let's get to it. – **Nyjal J. Drayton**

Introduction

AI is already shaping the world around us. The way we work, the way we create, the way we compete in business– AI is behind the scenes, making things faster, smarter, and more efficient. And yet, too many of us are still on the outside looking in, unsure of where we fit into this new reality.

Let's clear that up right now: **AI is for you.**

This book was written for the **DOERS, the builders, the visionaries**, the people who know that technology isn't just about innovation; it's also about opportunity. Whether you're a student, an entrepreneur, a creative, or someone looking to future-proof your career, **AI can be your biggest asset**–but only if you know how to use it.

What This Book Will Teach You

You don't need a computer science degree or a background in tech to make AI work for you. What you do need is **the right mindset, the right tools, and a clear understanding of how to apply AI in real ways.**
Here's what you'll learn:
- How to **use AI to work smarter, not harder**– automating repetitive tasks, improving productivity, and making better decisions.
- How AI can help you in your job search, career growth, and skill development–giving you a competitive edge in a rapidly changing job market.

- How entrepreneurs can leverage AI for market research, content creation, and business scaling–without needing a massive budget or a full team.
- The **darker side of AI**–including biases, misinformation, and ethical concerns–so you can navigate this space responsibly.
- The **right way to interact with AI**–because how you ask the right questions can be just as important as the answers you get.

Why This Matters Now

The world isn't waiting for us to catch up. AI is already in the hands of those who understand its power, and they're using it to **get ahead**. The real divide isn't between people who are "tech-savvy" and those who aren't–it's really between those who are **willing to adapt** and those who refuse to.

This book is your guide to making sure you're on the right side of that divide. No fluff, no unnecessary complexity–just real, actionable knowledge that you can start using today.

01

AI–What It Really Is

The Myth vs. Reality of AI

When most people hear "AI," they picture one of two things: either some futuristic robot ready to take over the world or a soulless machine replacing human jobs. But here's the truth–AI isn't some sci-fi fantasy, and it's not here to make us obsolete. AI is already woven into our daily lives in ways most people don't even notice.

That song recommendation on your playlist? AI.

The way your phone unlocks with your face? AI.

The "For You" page that somehow knows exactly what you like? AI.

Artificial intelligence has been in development for **decades,** but only recently has it evolved from a complex, industry-exclusive tool into something widely accessible. What was once confined to research labs and tech giants is now integrated into everyday life, empowering individuals, businesses, and creatives like never before.

Now, AI is *not* just for tech elites–it's for everybody. And if we don't learn how to use it, we risk being left behind while others get ahead. This chapter is about breaking AI down so it's clear, simple, and practical. No tech jargon, just real talk about how AI is shaping the world around us–and how we can make it work for *us*.

What AI Really Is (In Plain Terms)

At its core, AI is just a smart system that learns from patterns and data. It doesn't think for itself like a human, but it can analyze information at lightning speed and make predictions based on what it has seen before.

Think of AI is like an intern that learns over time. At first, it needs guidance, but the more information you feed it, the better it gets at understanding what you want.

AI is *not* magic–it's just math and algorithms working in the background to make things more efficient.
A few examples of AI at work today:

Voice Assistants - Siri, Alexa, Google Assistant

Streaming Services - Netflix suggesting your next binge

E-commerce - Amazon showing you exactly what you were *just* thinking about buying **Chatbots** - Customer service reps that aren't real people (but kind of sound like they are)

AI Art & Writing - ChatGPT, DALL·E, Midjourney (yes, this book is about that kind of AI too!)
AI is already shaping industries–from medicine to marketing, education to entertainment. The question isn't *if* AI will be part of our future. It's *how* we make sure our

communities aren't just consuming AI but actually using it to get ahead.

Why AI Matters for *Us*

For too long, emerging technology has been something that happens *to* us, not *for* us. We get the latest gadgets, but we aren't always in the rooms where the technology is created, shaped, or monetized. AI gives us a chance to change that.

Think about it like this: Instead of just watching content, you can create it faster and better.

Instead of waiting for job opportunities, you can use AI to level up your skills and make yourself more competitive.

Instead of working harder, you can work *smarter*–letting AI automate tasks and free up your time for bigger moves.

AI is like a toolbelt–it can't build your future for you, but it can give you the right tools to make the process easier, faster, and more effective. The people who take advantage of AI now will be the ones leading in the next decade. The ones who ignore it? They'll be playing catch-up.

The Role of Generative AI

While AI comes in many different forms, the kind that has taken center stage in recent years–and the type that is most accessible to us today–is **Generative AI**. Generative AI is a type of artificial intelligence that creates new content, whether that's text, images, music, or even videos. It doesn't just analyze and predict like other AI models; it actually **produces** new outputs based on the data it has been trained on.

Many of the AI tools referenced throughout this book, including tools like ChatGPT, and Midjourney, fall under this category. They generate new content based on user input, making them powerful tools for creativity, automation, and problem-solving.

However, AI is much broader than just Generative AI. Other types include:

- **Predictive AI** - Used in finance, healthcare, and business to forecast trends and make data-driven decisions.
- **Machine Learning AI** - The backbone of most AI models, allowing systems to improve over time based on experience.
- **Computer Vision AI** - Powers facial recognition, object detection, and even self-driving cars.
- **Conversational AI** - Chatbots and virtual assistants that simulate human-like interactions.

Generative AI is simply one piece of a much larger AI landscape. But because of its accessibility and ability to enhance our daily lives, it's the form of AI we'll be focusing on the most throughout this book. Understanding how to leverage it is key to making AI work **for us**, not against us.

02

Work Smarter, Not Harder– AI for Career Growth

The Old Way vs. The AI Way

For years, landing a job meant spending hours fine-tuning resumes, crafting cover letters from scratch, and sending out applications with the hope of a callback. Career growth felt like a grind.. One that required long nights, endless networking, and a bit of luck.

But today, AI is flipping the script. Instead of working harder, people are working *smarter*. You can use AI to optimize your resume, prep for interviews, and even find hidden job opportunities. You can automate parts of your workload, making room for career growth without burning out. The goal has turned from just *getting* a job, to *leveraging* AI to secure the best opportunities and stay ahead in an ever-changing job market.

AI-Powered Resumes & Cover Letters

A strong resume and cover letter can make or break your job application. But we can all probably agree that writing them from scratch is time consuming and exhausting. With AI-powered tools like ChatGPT, Resume Worded, and Teal

HQ, you can craft tailored resumes and cover letters in minutes.

Just imagine that for weeks, you've been sending out applications with no response. The rejection emails–or worse, the silence–are making you second-guess everything. After tweaking your resume over and over, you try an AI resume optimizer like Jobscan. It scans your resume, compares it to job descriptions, and pinpoints exactly what's missing–key terms recruiters are searching for, formatting adjustments, and ways to make your experience stand out. With a few quick fixes, your applications start getting more responses. A week later, you're scheduling interviews instead of sending out more applications.

AI tools help with resume optimization by analyzing job descriptions and suggesting key terms that increase visibility. They refine your wording to highlight your strengths, and if you struggle with writing cover letters, AI can generate a tailored draft based on the role you're applying for. Instead of spending hours staring at a half-blank page, you have a polished application ready to go in minutes.

Finding Jobs Faster with AI

Scrolling through endless job listings and applying one by one can feel like a full-time job itself. Another thing that AI can do is streamline that process by helping you find the best opportunities faster and even apply on your behalf.

AI job scouts help by finding relevant job openings and matching them to your skills. AI networking assistants suggest conversation starters and help you connect with industry professionals, while some tools even track applications and remind you when to follow up so nothing falls through the cracks.

Upskilling with AI-Driven Learning Platforms

The job market is shifting fast, and the skills that mattered yesterday might not be enough tomorrow. AI makes upskilling easier by personalizing learning experiences and helping you acquire in-demand skills quickly.

Instead of going back to school, you could explore AI-driven learning platforms like Khan Academy and Coursera. These platforms suggest structured learning paths based on what you already know, filling in the gaps with short, interactive lessons. AI tutors break down complex topics in a way that finally makes sense, and coding assistants help troubleshoot mistakes in real time. Within a few months, you could build a small portfolio and start applying to entry-level tech jobs.

AI-powered learning tools recommend courses based on your career goals, provide instant feedback on exercises, and help build real-world projects so you can showcase your skills.

But upskilling isn't necessarily about switching careers, unless that's what you want, it can also be about growing and expanding where you already are.

The Power of AI for Intrapreneurs

Not everyone wants to start a business, but that doesn't mean you can't think like an entrepreneur within your company. That's called being an *intrapreneur*–someone who innovates, leads, and builds new ideas from inside an organization.

AI can help you develop intrapreneurial skills by automating repetitive tasks, analyzing data for smarter decision-making, and even assisting in brainstorming creative solutions. If you can leverage AI to make processes more efficient, drive innovation, and bring value to your team, you make yourself an asset to any company.

Instead of just going through the motions at work, you could start incorporating AI to analyze sales trends and generate reports that help your team make better decisions. You'd be able to create a marketing copy in half the time using AI writing tools.

Being able to harness tools like these are one of the newest keys to making yourself indispensable in whatever space you're in.

Personal Branding with AI-Generated Content

Your online presence is often your first impression. No matter if you're job hunting, freelancing, or launching a business, AI can help you create content that attracts the right opportunities and elevates your professional brand.

AI-driven tools can refine and enhance your personal branding strategy in several ways. AI-powered writing assistants can help craft compelling LinkedIn bios, resumes, and portfolio descriptions that emphasize your strengths and achievements. Instead of spending valuable time fine-tuning your messaging, AI can generate polished, optimized content in seconds, to ensure your profile stands out to potential employers, clients, and collaborators.

For social media, AI tools can analyze engagement trends and suggest content ideas tailored to your industry, helping you maintain a consistent and impactful online presence.

These tools can recommend the best times to post, generate captions, and even provide insights into the type of content that resonates most with your audience. By leveraging AI for social media management, you can increase engagement, grow your network, and establish yourself as a thought leader in your field.

Beyond social media, AI-powered website builders, such as Wix ADI and other automated design platforms, can create professional portfolios and personal websites in minutes. These platforms streamline the process of designing, structuring, and optimizing a website, ensuring that your digital presence remains polished and effective.

Additionally, AI-driven SEO optimization tools can analyze search trends and suggest high-impact keywords to improve discoverability, making it easier for potential clients, recruiters, or business partners to find you online.

By integrating AI into your personal branding efforts, you can maximize visibility, credibility, and accessibility, positioning yourself for success in a competitive digital landscape.

The Bottom Line: AI Gives You the Edge

The workforce is evolving, and those who adapt will thrive. **AI won't do the work for you**, but it can make your job search, skill-building, and personal branding *a whole lot easier*. The key is knowing how to use it effectively.

The people who master AI in their careers today will be the ones leading tomorrow.

So let's get ahead of the game. This book is about *making AI work for us*—not just learning what it is, but actually using it to build, create, and secure the bag.

03

The Hustler's Guide to AI– Using Tech to Make Money

Why AI is the Ultimate Sidekick for Entrepreneurs

Entrepreneurship has always been about solving problems, finding opportunities, and making things happen. The reality is, running a business–whether full-time or as a side hustle–can be overwhelming. There are only so many hours in a day, and for most people, managing everything from customer service to marketing to financial tracking can feel like a never-ending grind.

This is where AI can step in. It'll never be about replacing human effort, but about making the process more efficient. AI acts as a *force multiplier*, automating tasks that eat up time, providing insights that would take hours to uncover manually, and helping businesses stay competitive in a fast-moving digital world.

AI for Market Research & Finding the Right Customers

Understanding your audience is the foundation of any successful business. If you don't know who your customers are, what they need, and how they make decisions, you're basically shooting in the dark. Traditional market research often involves time-consuming surveys, expensive focus groups, and sifting through boatloads of data manually. AI changes this by providing instant access to insights that would normally take weeks, sometimes months to gather.

AI-powered tools can analyze things like online behavior, purchasing trends, and social media interactions to help businesses understand what their target audience is interested in. Tools like **Google Trends** and **AnswerThePublic** show what people are searching for in real time, helping entrepreneurs identify market demand. More advanced AI, like ChatGPT, can analyze customer feedback and summarize common themes, making it easier to adjust products and services based on actual needs.

For those selling products, AI-driven platforms like **Shopify's AI analytics** can identify which items are gaining traction, helping businesses focus on their best sellers. Even social media platforms like **Meta's Audience Insights** use AI to help advertisers understand which demographics are most likely to engage with their content.

I'd say if you're new to AI, start by using Google Trends to explore industry trends. Look up a keyword related to your business and see how interest changes over time. If you're a bit more experienced, AI-driven customer segmentation tools like **CrystalKnows** or **HubSpot AI** provide deeper insights into consumer behavior, helping refine marketing strategies.

Automating Customer Service & Engagement

Customer service can make or break a business. A great product isn't enough if customers feel ignored, frustrated, or undervalued. Traditionally, businesses either have to handle customer inquiries themselves (Which is just time-consuming and quite frankly, no one wants to sit around doing) or hire a support team (which is expensive). AI offers a third option: **automation**.

AI chatbots like **Tidio, ManyChat, and Drift** allow businesses to offer 24/7 customer support without human involvement. These bots can answer frequently asked questions, process basic requests like order tracking, and even schedule appointments.

More advanced AI tools integrate with messaging apps like **WhatsApp and Facebook Messenger**, allowing businesses to automate conversations on the platforms customers are already using. Instead of waiting on hold or sending an email, customers get instant responses.

For e-commerce businesses, AI-driven chatbots can also recommend products based on browsing behavior, turning customer service into a sales tool.

Use a free chatbot tool like Tidio to set up automated responses for common questions (shipping times, refund policies, etc.). Or for more experienced users, AI-powered sentiment analysis tools like **MonkeyLearn** analyze customer feedback in real time, identifying patterns in complaints or praises, allowing businesses to adjust accordingly.

AI Tools for Content Creation & Marketing

Content marketing is one of the most effective ways to grow a brand, but creating high-quality content consistently remains a challenge. AI-powered tools are now helping to streamline this process by generating blog posts, captions, graphics, videos, and even voiceovers in a fraction of the time it would normally take.

AI-powered text generators like **ChatGPT, Jasper AI, and Copy.ai** help businesses create engaging social media posts, product descriptions, and ad copy. Instead of struggling with writer's block, which can sometimes last a while for some writers, AI can generate content ideas, improve messaging, and even suggest hashtags to boost engagement.

For visual content, platforms like **Canva's AI tools and Midjourney,** which is one of my favorites, generate high-quality graphics in seconds. Businesses who adopt these skills and techniques internally using these types of tools no longer need to rely on expensive designers for every social media post or marketing campaign.

Even video marketing has been revolutionized by AI. **Eleven Labs** generates human-like voiceovers, while tools like **Synthesia** create AI-generated video content, allowing businesses to create professional videos without hiring actors or expensive production teams.

Try things like using ChatGPT to generate a week's worth of social media captions for your business in minutes. Or using AI-driven video editing software like **Descript** to automatically transcribe, edit, and or create content from raw video footage, saving hours of manual work.

Scaling Without a Huge Team

One of the biggest barriers to growth is operational capacity. As a business grows, so does the workload–more emails, more orders, more financial tracking. Traditionally, scaling means hiring more people, but in this new generation AI allows businesses to grow while keeping costs low.

AI automation tools like **Zapier and Notion AI** help businesses connect different apps and automate mundane tasks. For example, instead of manually sending invoices, an AI-powered system can automatically generate and send them when a purchase is made.

AI accounting tools like **QuickBooks AI and Xero** analyze financial transactions, track expenses, and even identify areas where costs can be reduced.

AI inventory management tools are made to help businesses, predict demand and optimize stock levels, preventing over-ordering or running out of products.

The Bottom Line: AI Makes the Hustle Smarter

Entrepreneurship is still about creativity, strategy, and execution–but AI is making it easier than ever to compete and grow. Instead of getting stuck in the day-to-day grind of manual tasks, AI allows business owners to focus on big-picture strategies, customer relationships, and long-term success. Those who learn how to integrate AI into their hustle will have a serious edge.

04

The Creative's Blueprint—AI for Content & Culture

How AI is Reshaping Creativity

Creativity has always been our superpower. From music and art to fashion and storytelling, culture has been our canvas. Now, AI offers new tools to preserve, express, and amplify our culture in ways that were once unimaginable.

For some, AI has become a game-changer, making creative tools more accessible than ever. For others, it raises concerns about originality and authenticity, which are fair concerns to have. But one thing is clear—AI is not replacing creativity, nor will it ever. It's merely amplifying it. It's giving artists, designers, and storytellers a new kind of paintbrush, one that can turn imagination into reality faster than ever before.

This thing is bigger than just automating tasks— this is about opening up new possibilities. Imagine using AI to recreate the sounds of cultural legacies, design digital art that tells our stories, or craft narratives that reflect the richness of our experiences. This technology can preserve traditions while also inspiring new forms of expression.

AI in Music & Audio Production

AI is changing the way music is made, from beat production to mastering. Musicians and producers can now use AI-powered platforms to generate melodies, remix tracks, and even create entirely new compositions with just a few clicks.

AI-powered tools like **Boomy and AIVA** help users compose original music in seconds. These platforms analyze patterns from existing songs and create instrumentals based on different styles, something even the most experienced producers struggle with. Whether you're a seasoned producer or just experimenting, AI can be a collaborator, generating ideas you might not have thought of on your own. Although there is truly nothing new under the sun, there's no way you can think of everything.

Beyond composition, AI is also streamlining **sound editing and mixing**. Platforms like **Landr** use AI to master tracks instantly, removing the need for expensive studio sessions. AI voice generation tools like **Eleven Labs** also make it easier to create voiceovers or alter vocals without hiring an artist.

For independent creators and content producers, this means music production is no longer limited to those with big budgets or studio access– essentially AI is democratizing this process.

AI-Generated Visual Art & Design

Let's look at the art industry and how AI-generated art has sparked both excitement and controversy in the creative world. Just like music and audio production, some see it as a revolutionary tool that expands creative possibilities, while others, again, worry about originality and the role of human artists. The reality? AI is best used as a *tool* rather than a replacement.

AI-powered design tools like **DALL·E, Midjourney, and Runway ML** allow artists and designers to generate high-quality images from simple text descriptions or *prompts*. These platforms can create anything from surreal digital paintings to realistic product mockups in seconds.

Graphic designers can use AI for **speeding up workflows**, creating concepts, and enhancing creativity. Tools like **Adobe Firefly** assist with generating design elements, while **Canva AI** makes branding and content creation more efficient for businesses. Instead of replacing human designers, AI acts as an assistant, automating repetitive design tasks like background removal, color correction, and resizing.

For content creators, AI-generated visuals mean **faster content production**. Now, when doing things like designing a book cover, a social media campaign, or an animation, AI can give you a starting point that's customized to match your vision.

AI & Writing: From Content Creation to Storytelling

AI is transforming how writers generate ideas, structure content, and refine their work. AI writing tools help with everything from **brainstorming and drafting** to **editing and language translation**.

Platforms like **ChatGPT, Jasper AI, and Sudowrite** assist with writing blog posts, ad copy, and long-form storytelling. For novelists, screenwriters, and journalists, AI can generate outlines, suggest character arcs, and improve flow without taking away the human voice.

Editing and grammar-checking tools like **Grammarly and Hemingway Editor** have been AI-powered for years, helping writers refine their work with precision. AI also plays a major role in **speech-to-text transcription**, making it easier for podcasters, interviewers, and journalists to turn spoken words into written content.

While AI is a powerful assistant, storytelling remains deeply personal. The best use of AI in writing is **to enhance the creative process, not replace it**. Writers still bring the emotion, the voice, and the originality–AI simply helps execute the vision faster.

The Balance Between AI & Originality

As AI becomes more involved in creative fields, the question of originality is becoming more important. If an AI tool generates an image, a song, or a story based on learned patterns, is it truly original? More importantly, who owns it?

Artists and creators are beginning to define their own approach to AI. Some use it as **a starting point**, refining AI-generated content to add their personal touch. Others integrate AI into their workflow but ensure that their human creativity remains the driving force.

The best way to look at AI in creativity is this: it's another tool in the creative toolbox. Just as digital software revolutionized photography and music production, AI is simply the next evolution. The creators who learn to use it effectively will have an advantage, but **true originality will always come from the human mind**.

The Bottom Line: AI Expands Creative Possibilities

AI is transforming how we create, but it's not here to take away artistic vision–it's here to expand it. You could be an artist, a musician, a writer, or a content creator. AI can make the creative process smoother, faster, and more experimental for anyone. When we use AI to elevate our cultural expression, we're not just keeping up with technology–we're setting the pace, showing the world how culture and innovation come together.

05

AI & the Bag–Money Management & Wealth Building

AI is Changing the Way We Handle Money

Money management has always been about making smart decisions–spending wisely, saving consistently, and investing for the future. But in today's world, financial literacy alone isn't always enough, especially in a lot of our communities. With AI stepping into the financial space, people now have access to tools that analyze spending habits, predict market trends, and automate wealth-building strategies.

AI is helping to reshape personal finance. If you're looking to budget smarter, build an investment portfolio, or protect yourself from fraud, AI is making financial management easier, faster, and more accessible than ever before, and it's only getting faster, and more efficient.

Smarter Budgeting with AI

Traditional budgeting often means manually tracking expenses, categorizing spending, and adjusting financial goals over time. But AI-driven budgeting tools take the guesswork out of money management by analyzing spending patterns, forecasting future expenses, and providing personalized financial insights.

AI-powered apps can track transactions in real time, identify areas where money is being wasted and suggest ways to save. Instead of relying on fixed budgeting templates, these tools create a flexible plan that adjusts based on income, spending habits, and financial goals.

For those who struggle with saving, AI tools automate the process. Some tools round up spare change from purchases and transfer it into savings, while others allocate money into different categories based on spending trends. The result? A smarter approach to budgeting without the constant number-crunching.

AI & Investing: Making the Market More Accessible

Investing has always been one of the most effective ways to build wealth, but for many people, the stock market feels intimidating. AI is lowering the barrier to entry by providing automated investment strategies, real-time market analysis, and risk assessment tools.

Robo-advisors use AI to create personalized investment portfolios based on individual financial goals, risk tolerance, and market trends. Instead of manually researching stocks or relying on financial advisors, AI analyzes thousands of data points to optimize investment decisions.

AI is also enhancing **market forecasting**. While no tool can predict the market with 100% accuracy, AI-powered financial models process historical trends, global events, and economic indicators to provide smarter insights on when to buy, sell, or hold.

For beginners, AI investment platforms offer guidance on where to start, while experienced investors use AI-driven analysis to refine their strategies and identify new opportunities.

Fraud Detection & AI-Powered Financial Security

As digital transactions become more common, so do financial scams and identity theft. AI is stepping in to play a crucial role in fraud detection by analyzing spending behaviors and identifying suspicious activity in real time.

Banks and financial institutions use have been using AI for years to track transaction history and flag anything unusual–like purchases in different locations or unusually large withdrawals. Many fraud detection systems now use machine learning to recognize patterns in cybercrime, helping to prevent unauthorized transactions before they happen.

On a personal level, AI-powered security tools monitor credit reports, detect data breaches, and provide alerts if financial information is compromised. As AI continues to evolve, its role in financial security will only become more essential in protecting personal assets.

Building Long-Term Wealth with AI

Wealth-building isn't all about earning more money, it's truly about making smart financial decisions over time. AI tools helps with long-term financial planning by offering insights on savings goals, retirement strategies, and investment opportunities tailored to individual circumstances.

AI-driven retirement planning tools calculate how much money is needed for future financial goals and adjust strategies based on changes in income and expenses. Instead of relying on traditional financial planning methods, AI adapts in real time, making adjustments as market conditions and personal situations evolve.

The key to long-term financial success is *consistency*, and AI automates many of the processes that people struggle with. It could be setting up recurring investments, reallocating assets, or optimizing tax strategies. The more AI is integrated into financial planning, the easier it could become to stay on track and build lasting wealth.

The Bottom Line: AI is the Financial Cheat Code

Managing money has always required discipline and strategy, but AI is making it easier to stay in control. From automated budgeting and AI-powered investing to fraud prevention and long-term wealth planning, technology is giving people more financial power than ever before.

But with these advancements come new risks–bias, misinformation, and automation replacing jobs. Before we go any further, let's take a step back and explore the challenges AI presents, so we can navigate them wisely.

06

The Dark Side of AI—Risks, Biases & Unintended Consequences

AI is Powerful—But It's Not Perfect

For all the ways AI is changing the world for the better, it also comes with **risks, challenges, and unintended consequences**. The more AI becomes integrated into daily life, the more we need to be aware of **its flaws, biases, and potential for misinformation**.

AI doesn't think, feel, or understand context the way humans do. It processes data, identifies patterns, and makes predictions—but **it doesn't always get things right**. When used irresponsibly or without oversight, AI can spread misinformation, reinforce biases, and even make decisions that negatively impact real people.

This chapter breaks down **the risks of AI,** how to spot them, and what people need to do to **stay informed and protected** in an AI-driven world.

AI Bias: When the Data is Flawed, So Are the Results

AI is only as good as the data it's trained on. If that data is **incomplete, unbalanced, or biased**, then AI models will **reflect and amplify those biases**, often without users realizing it.

- **Hiring Discrimination** - AI-powered hiring tools have been shown to favor certain demographics over others because they were trained on biased hiring data from the past.
- **Facial Recognition Flaws** - Some AI-powered facial recognition systems struggle to accurately identify people with darker skin tones, leading to wrongful arrests and false matches.
- **Healthcare Inequality** - AI-driven medical tools trained on **limited or unrepresentative patient data** have made incorrect diagnoses, disproportionately affecting underserved communities.

But AI doesn't create these biases on its own, it's **learning it from human-generated data**. The problem is that AI can **scale bias at an unprecedented level**, making it more difficult to detect and correct.

How to Navigate AI Bias:

- **Be skeptical of AI-driven decisions**, especially in hiring, policing, and finance.
- **Push for transparency**–AI should be explainable, and its training data should be scrutinized.

- **Demand diversity in AI development**–the people building AI should represent the full spectrum of society, not just a select few.

AI Models Are Like Raising a Child–Why Representation Matters

AI is something that learns, evolves, and adapts based on what it's exposed to. In a lot of ways, training an AI model is like raising a child. The way it's taught, the data it's given, and the values embedded in it during its early stages will shape how it behaves as it grows. Just like a child absorbing language and social norms from their environment, AI models learn from the data they are trained on. If that data is diverse and balanced, the AI will reflect a wider range of experiences, perspectives, and cultures. But if the data is biased, limited, or skewed toward one demographic, the AI will reflect those same gaps–sometimes in ways that reinforce stereotypes and exclude certain communities. As AI advances, it starts to develop more autonomy–much like a teenager gaining independence. This phase is crucial because if the AI wasn't trained responsibly, it may generate misinformation, amplify biases, or make flawed decisions. Without strong ethical oversight, AI can "hallucinate" false facts, exclude marginalized voices, or perpetuate inequalities simply because it wasn't trained with representation in mind. When AI reaches its "adulthood" and becomes deeply embedded in our daily lives, its influence will be everywhere–from hiring decisions to legal systems, from healthcare to finance. By this stage, if we haven't ensured representation in AI's development, we'll be dealing with **a world where**

technology is built for only a select few, instead of for everyone.

This is why representation in AI development isn't just important—it's **urgent**. If diverse voices aren't part of shaping AI models today, then when AI reaches its full potential, it will leave entire communities out of the future.

The way we train AI now determines how fair, just, and inclusive it will be tomorrow. If we want AI that represents **all of us**, we need to be in the room, shaping how it learns today.

AI Hallucinations: When AI Makes Things Up

One of the biggest weaknesses of AI-powered chatbots like ChatGPT is their tendency to **hallucinate**—a term used when AI **confidently presents false or misleading information as fact.**

Because AI doesn't "know" things the way humans do, it generates responses based on probabilities. That means it can **misinterpret facts, fabricate details, or create sources that don't exist.**

For example:

- AI-generated research papers have included **fake citations**—references to academic sources that don't actually exist.
- Some AI chatbots have created **historically inaccurate timelines,** mixing real events with fabricated ones.

- AI-generated news articles have **spread misinformation** because they misinterpret real-world events.

How to Spot & Correct AI Hallucinations:

- **Always fact-check AI-generated content–** especially for research, news, and legal/financial advice.
- **Ask for sources–**if an AI-generated response includes citations, verify that those sources actually exist.
- **Use AI as an assistant, not the final authority–** it should enhance human intelligence, not replace it.

AI & Misinformation: Fake News at Scale

The rise of AI-generated content has made **misinformation easier to produce and harder to detect**. AI-powered deepfakes, synthetic media, and chatbot-generated false narratives can **spread quickly and influence public opinion** before the truth catches up.

- **Deepfakes** - AI-generated videos that manipulate real footage, making it appear that someone said or did something they never did.
- **AI-Generated Fake News** - AI-written articles designed to mislead, whether for political manipulation, financial scams, or social engineering.
- **Social Media Bots** - AI-driven fake accounts that amplify false information, creating the illusion of widespread support or controversy.

Misinformation isn't just a small inconvenience, it holds the potential to **impact elections, spread conspiracy theories, and even incite violence**. The more sophisticated AI becomes, the harder it will be to distinguish between **what's real and what's artificially generated**.

How to Defend Against AI-Generated Misinformation:

- **Verify before sharing**–check multiple sources before believing sensational headlines.
- **Use AI detection tools**–platforms like **Deepware and Sensity AI** help identify AI-generated deepfakes.
- **Follow credible sources**–AI-generated content thrives on **low-quality, unverified platforms**, so stick to trusted journalism and fact-checking sites.

Privacy & Surveillance: How AI is Tracking You

AI is always **collecting and analyzing personal data at an unprecedented scale** and influencing what we see online. Every interaction with AI-powered tools, search engines, and smart devices leaves a digital footprint that can be **tracked, stored, and monetized**.

- **AI in Social Media** - Platforms use AI to track user behavior, predict interests, and serve highly targeted ads, sometimes **manipulating emotions and engagement**.

- **AI-Powered Surveillance** - Governments and corporations are using AI-powered facial recognition and predictive analytics to **monitor people's movements, online activity, and personal habits.**
- **Data Leaks & Cyber Threats** - AI-driven hacking tools can analyze vulnerabilities faster than human hackers, making cybersecurity a growing concern.

How to Protect Your Privacy in the Age of AI:

- **Limit data sharing**–adjust privacy settings on apps and devices.
- **Use encrypted communication tools**–like Signal or ProtonMail for sensitive messages.
- **Be cautious with AI assistants**–voice-activated devices like Alexa and Google Assistant **are always listening**, literally always, so review privacy settings carefully.

AI & Job Displacement: Who Gets Left Behind?

While AI is creating new job opportunities, it's also eliminating roles that rely on repetitive, predictable tasks. Many industries are feeling the shift:

- **Customer Service** - AI chatbots are replacing human support agents for handling basic inquiries.
- **Retail & Warehousing** - AI-powered automation is reducing the need for human workers in logistics and inventory management.
- **Administrative Roles** - AI-driven scheduling, bookkeeping, and data entry are cutting traditional office jobs.

For those in **at-risk careers, upskilling and adapting to AI-powered work environments is crucial**. The people who succeed in an AI-driven world will be those who learn **how to work with AI rather than *try* to compete against it**.

How to Stay Ahead of AI-Driven Job Changes:

- **Develop skills that AI can't easily replace**, like creative problem-solving, emotional intelligence, and leadership.
- **Learn how to use AI as a tool**—mastering AI-driven platforms will make workers more valuable rather than obsolete.
- **Explore careers in AI-related fields**, such as AI ethics, human-AI collaboration, and AI security.

The Bottom Line: AI is Powerful—But Needs Oversight

AI is not inherently good or bad—it's a tool. How it impacts society depends on how we develop, regulate, and use it. The risks of AI—bias, misinformation, hallucinations, privacy violations, and job displacement—are real, but they can be managed if people stay informed and proactive.

The key takeaway? AI is not infallible. Fact-check everything, question AI-driven decisions, and never assume AI-generated content is always correct. The more we understand the risks, the better equipped we are to use AI responsibly and hold its creators accountable.

Now that we understand the challenges AI presents, it's time to look forward. The future of AI isn't just about risk– it's also about opportunity. AI is evolving rapidly across industries, changing how we work, create, and compete. The question isn't whether AI will be part of our future, but how we can ensure it works for us, rather than against us.

07

The Future is Ours–What's Next in AI?

AI is Just Getting Started

AI is evolving fast, and the stakes are high. As technology reshapes industries and economies, the risk isn't just about being left behind in the job market. It's about being excluded from the spaces where AI is developed, shaped, and monetized. If we're not represented in those rooms, decisions about how AI impacts culture, community, and opportunity will be made without us.

We've led every wave of innovation in culture, from shaping global music trends to defining fashion and digital movements. The next wave is AI, and it's already here. But the question is: will we be creators in this space, or consumers?

From healthcare to education, entertainment to entrepreneurship, AI is becoming deeply embedded in industries that impact daily life. While some people worry about what AI means for job security and ethics, others are embracing it as an opportunity to build new skills, create new businesses, and shape the future instead of reacting to it.

The Expanding Role of AI Across Industries

The impact AI has is already spanning across nearly every field, and its influence is only growing. Some industries are being disrupted completely, while others are using AI to enhance efficiency and open new opportunities.

- **Healthcare** - AI-powered diagnostics, robotic-assisted surgeries, and predictive medicine are transforming patient care. AI is helping doctors detect diseases earlier and create personalized treatment plans based on genetic data.
- **Education** - AI tutors, adaptive learning platforms, and AI-generated lesson plans are making education more personalized, allowing students to learn at their own pace.
- **Entertainment** - AI is being used to create music, scripts, and even full-length films. Streaming platforms use AI to personalize recommendations, and AI-generated influencers are gaining real audiences.
- **E-Commerce & Retail** - AI-driven recommendation engines, personalized marketing, and automated supply chain management are helping businesses provide better shopping experiences while maximizing efficiency.
- **Finance & Banking** - AI-powered fraud detection, robo-advisors, and automated financial planning tools are giving people more control over their money.

The industries that embrace AI early will be the ones that thrive, while those that resist will struggle to keep up.

The Bigger Picture: How AI is Reshaping Society

As we are in this time where AI is reshaping entire industries, economies, and the very fabric of society. We often focus on **how AI can benefit us personally**, but it's just as important to recognize the larger shifts AI is driving on a global scale.

AI is Rewiring the Economy

- The job market is shifting. Automation is replacing traditional roles while creating **entirely new fields of work**. AI-driven industries—such as **autonomous vehicles, personalized healthcare, and algorithmic finance**—are becoming billion-dollar sectors. Countries that invest in AI development are setting themselves up for economic dominance, while those that lag behind risk being left out of the next industrial revolution.

AI is Reshaping Infrastructure & Urban Life

Smart cities powered by AI are already in development, using machine learning to optimize **traffic systems, energy consumption, and public services**. AI-driven surveillance, while controversial, is being used in urban planning, law enforcement, and even disaster response. As AI continues to evolve, its presence in daily life will be as fundamental as electricity or the internet.

AI in Politics & Global Power Struggles

AI is also a **geopolitical force**, with nations competing to lead in AI research, cybersecurity, and military applications. From AI-powered disinformation campaigns to deepfake technology influencing elections, governments are actively exploring both the power and risks of AI. The countries and companies that dominate AI development today **will shape the policies, economies, and digital ecosystems of tomorrow**.

This rapid growth and development of AI is changing the world. And if we don't claim a seat at the table now, or even better, build our *own* table, we risk becoming spectators in a game where others define the rules.

Big Tech's Billion-Dollar AI Race: What It Means for Us

AI isn't just changing how we work, but also triggering one of the biggest investment waves in tech history. The world's biggest companies are pouring **billions** into AI research, development, and infrastructure, pushing the limits of what's possible.

Project Stargate & the Future of Super-intelligent AI

Tech giants are racing to build the most advanced AI models the world has ever seen. One of the most ambitious projects, **Project Stargate**, is rumored to be a large-scale $500 billion AI initiative focused on developing **massively scalable, next-gen AI systems**. The details are still emerging, but insiders suggest that it could redefine how AI interacts with the world, pushing us closer to AGI (Artificial General Intelligence).

Meta's AI Hubs & the Expansion of AI Infrastructure

Meta (formerly Facebook) is investing heavily in **AI research hubs** around the world, planning to build massive AI-focused data centers. These hubs will serve as training grounds for advanced AI models, improving areas like **computer vision, natural language processing, and personalized recommendations**. The goal? To create an ecosystem where AI can interact with users more intuitively than ever.

Tech Giants' AI War: A Trillion-Dollar Market Shift

Companies like **Google, Microsoft, Amazon, and OpenAI** are in an AI arms race, each investing billions to outpace competitors. Microsoft alone has invested over **$10 billion in OpenAI**, while Google is doubling down on AI-powered search and assistants. These investments show that AI isn't just a trend—it's becoming the **backbone of the future economy**.

So, what does this mean for us?

It signals a shift where AI will become as essential as the internet itself. The companies controlling AI infrastructure today will shape the job market, business landscape, and even how we interact with technology. Instead of being passive consumers, we need to position ourselves as **creators, entrepreneurs, and innovators who understand and leverage AI**. The AI revolution is here!

The question now is, how will we make sure our communities aren't just watching from the sidelines?

AI and Global Power: The Race for Control

Artificial intelligence isn't just a tool for businesses and individuals—it's becoming a defining force in **global power dynamics**. Governments, corporations, and policymakers are locked in a **battle for AI dominance**, knowing that those who control AI technology today will shape the world of tomorrow.

The AI Arms Race: Who Controls the Technology?

Nations around the world are investing billions into AI research, competing for technological supremacy. The United States, China, and the European Union are leading the charge, each developing AI models, infrastructure, and regulatory frameworks that will define the next industrial revolution.

China's AI Expansion - China has made AI development a national priority, integrating it into military defense, mass surveillance, and economic policy. The government has invested in AI-driven smart cities and facial recognition systems, using technology to **control data and influence public behavior.**

The United States and Private Sector Dominance - Unlike China's government-led AI strategy, the U.S. relies heavily on private companies such as OpenAI, Google, and Microsoft. These companies are shaping AI's trajectory, raising concerns about whether AI development should be regulated more strictly.

The EU's Ethical AI Approach - The European Union is leading the charge in **AI regulation**, ensuring ethical considerations are built into AI development. With policies like the **AI Act,** the EU aims to prevent AI misuse while fostering innovation.

The question remains: Should AI be **regulated like a public utility** to prevent monopolization, or should it remain a **free-market competition** where the most advanced developers dictate the rules?

The Battle Over Data: Who Owns Information?

AI is **only as powerful as the data it's trained on**, and right now, that data is controlled by a handful of governments and corporations. Data privacy has become a global issue, with policies such as **Europe's GDPR and debates in the U.S. over AI-driven surveillance**.

Big Tech's Role in AI Governance - Companies like Google, Microsoft, and Meta hold vast amounts of data, raising concerns about AI bias, misinformation, and corporate influence over public discourse.

AI and National Security - Intelligence agencies worldwide are using AI for cybersecurity, counterterrorism, and deepfake detection. While this strengthens national security, it also introduces ethical concerns about government surveillance and data privacy.

The Rise of AI Regulations - Countries are moving to regulate AI before it spirals out of control. From the U.S. Senate hearings on AI ethics to China's strict content moderation policies, the world is trying to balance **innovation with accountability**.

So Who Writes the Rules?

AI is still in its early stages, but **whoever sets the policies today will control the future**. The decisions made by governments and corporations will shape:

- Whether AI remains accessible or becomes centralized in the hands of a few.
- How much influence AI has over job markets, political systems, and economic structures.
- What ethical considerations are built into AI to prevent harm and bias.

This is why it's crucial for **diverse voices and perspectives** to be included in AI governance. AI should not be built solely for the benefit of **a select few nations or corporations**–it must serve global communities equitably.

The future of AI regulation isn't simply about the technology, but about **power, ethics, and the kind of world we want to build**.

AI & Job Security: Threat or Opportunity?

Again, one of the biggest concerns about AI is whether it will take away jobs. The reality is, AI is automating certain tasks, but it's also creating entirely new roles and opportunities. The key is learning how to work with AI rather than being replaced by it.

Jobs that involve **routine, repetitive tasks** are at the highest risk of automation. This includes roles in data entry, customer service, and manufacturing. However, jobs that require **critical thinking, creativity, and emotional intelligence** are much harder for AI to replicate.

Instead of fearing automation, the best move is to **develop AI-related skills** that will be in demand in the future. Learning how to use AI tools effectively, understanding data analysis, and improving digital literacy will make people more valuable in the workforce.

Once again, the biggest winners in the AI era will be those who learn to **use AI to increase their productivity** instead of trying to ignore it.

Ethical Considerations & AI Bias

As powerful as AI is, it's not perfect. As mentioned in chapter 6, one of the biggest challenges in AI development is **bias**–the fact that AI models can sometimes reflect and even amplify societal inequalities.

Because AI learns from existing data, it can pick up biases present in that data. This has been seen in hiring algorithms that favor certain demographics, facial recognition technology that struggles with accuracy across different skin tones, and AI-driven decision-making systems that unintentionally discriminate.

Addressing AI bias requires **greater transparency and diversity in AI development**. The more diverse the teams building AI systems, the less likely these biases will be built into the technology.

On a broader scale, **AI ethics** is becoming a major topic as governments and organizations set regulations for how AI should be used. Privacy concerns, data security, and AI accountability are all areas that will shape how AI is deployed in the future.

While these issues are real, they also present an opportunity for people to **get involved in shaping the future of AI**–whether that's advocating for ethical AI use, working in AI ethics research, or simply staying informed about how AI is being used.

How to Stay Ahead in the AI Era

The people who succeed in an AI-driven world won't be the ones who know the most about AI technology itself–it will be the ones who know how to **apply AI effectively** to their careers, businesses, and creative work.

Here's how to stay ahead:

- **Stay Curious** - AI is evolving rapidly. Keeping up with trends and new tools will give you an edge in adapting to changes.

- **Develop AI Skills** - You don't need to be a programmer to benefit from AI. Learning how to use AI-driven platforms in your industry will keep you competitive.
- **Experiment & Adapt** - The best way to understand AI is to **use it**. Whether it's for content creation, automation, or business growth, experimenting with AI tools will help you see where it fits into your workflow.
- **Think Bigger** - AI is a tool, but what you do with it is what matters. Those who see AI as an opportunity rather than a threat will be the ones shaping the future instead of being left behind.

The Bottom Line: AI is Here

The future of AI and how will shaped is cultural. It's about ensuring that our experiences are not only represented but celebrated. Now is the time to lead, build, and innovate–because if we don't, someone else will.

08

The AI Playbook–How to Communicate & Work Smarter with AI

AI is Only as Good as the Questions You Ask

One of the biggest misconceptions about AI is that it just knows everything. In reality, AI is only as good as the input it receives–the way you communicate with AI directly impacts the quality of the response you get.

This is where prompt engineering comes in. Learning how to structure your questions, provide the right amount of context, and refine responses makes AI a powerful tool for problem-solving, creativity, and efficiency.

Whether you are using AI for content creation, research, business strategy, or automation, understanding how to communicate effectively with AI will separate those who use AI casually from those who leverage it to its full potential.

What is Prompt Engineering?

Prompt engineering is the skill of crafting clear, precise, and effective instructions that guide AI to generate the best possible response. Instead of treating AI like a search engine, like Google, where vague keywords are enough, AI requires structured input to deliver high-quality results.

Think of it like giving directions—the more specific and clear you are, the better the outcome. A poorly written prompt can lead to vague, inaccurate, or even misleading AI responses. A well-structured prompt guides AI toward the most relevant and useful answer.

Example of a Weak Prompt: "Tell me about business." This is too vague. The AI does not know if you mean business strategy, starting a business, business history, or something else entirely.

Example of a Strong Prompt: "Explain the fundamentals of a business model canvas and how entrepreneurs can use it to structure a startup." This gives AI clear direction, specificity, and a defined scope, leading to a much more useful response.

The Key Elements of an Effective AI Prompt

A strong prompt typically includes:
- **Clarity** - Be direct about what you want (e.g., "Generate a step-by-step guide on..." instead of just "Help me with...").

- **Context** - Provide background details (e.g., "Explain machine learning for someone with no tech experience.").
- **Format** - If you need AI to generate something in a specific way, tell it (e.g., "Write this in a list format" or "Summarize this in two paragraphs").
- **Tone & Audience** - AI can adjust tone based on your needs (e.g., "Write in a formal tone for a corporate audience" vs. "Make this fun and engaging for beginners").

The more specific and well-structured your prompt is, the better AI can tailor its response.

Advanced Prompt Techniques for Maximum Efficiency

Once you understand the basics, you can start refining your prompts to get even more precise results.

A. Iterative Refinement

Instead of settling for AI's first response, tweak and refine prompts to improve the output.

- **First Prompt:** "Give me tips on marketing."
- **Refined Prompt:** "Provide five social media marketing strategies for a startup with a low budget."
- **Even Better Prompt:** "Generate five social media marketing strategies for a startup with less than $5,000 in ad spend, focusing on organic engagement and influencer partnerships."

Each refinement guides AI closer to the exact insights you need.

B. Role-Based Prompting

You can instruct AI to respond as if it were an expert in a specific field.

- **Example:** "Act as a marketing consultant and create a 90-day content strategy for a new e-commerce brand targeting Gen Z consumers."
- **Why it Works:** By assigning AI a role, it frames its response from a specialized perspective, making the answer more useful.

C. Multi-Step Prompts for Complex Tasks

Instead of asking for one big response, break tasks into steps.
- **Example:** "First, outline a business idea based on emerging AI trends. Then, create a five-step strategy to validate the idea. Finally, suggest the best platforms for reaching potential investors."
- **Why it Works:** AI will follow a structured process instead of giving a generic response.

Building Custom AI for Specialized Tasks

AI is more than just a tool for general productivity–it can be **customized** to handle highly specific tasks across industries. By tailoring AI to fit individual needs, professionals can automate complex workflows, reduce inefficiencies, and enhance their decision-making.

However, while AI offers significant benefits, its limitations —especially in areas like hiring and healthcare—must be acknowledged. In these fields, AI should be used as a **supporting tool**, not an **unquestioned** decision-maker.

Education & Learning

Lesson Planning - Educators can use AI to generate structured lesson plans based on curriculum guidelines, student learning objectives, and assessment needs. Instead of spending hours structuring content, AI can create a framework that teachers can refine and personalize.

AI-Powered Personalized Tutoring - AI-powered tutors can provide individualized learning assistance by identifying student weaknesses and tailoring explanations accordingly. These systems can analyze quiz responses, highlight weak areas, and adjust explanations in real-time to reinforce learning. Adaptive learning platforms can also create dynamic learning paths based on student progress, ensuring personalized education.

Business & Legal

Contract Analysis - AI can scan legal documents, highlight key clauses, and flag potential risks such as hidden fees or unfavorable terms. These systems analyze legal terminology, identify inconsistencies across documents, and suggest revisions for improved clarity and compliance. While AI speeds up contract review, legal professionals should always verify AI-generated insights for accuracy.

AI for Proposal & Grant Writing - AI can assist in drafting structured proposals for funding, grants, and business investments by aligning responses with specific requirements. While AI-generated drafts provide a starting point, human refinement is essential to maintain strategic alignment and authenticity.

Finance & Accounting

Financial Audits - AI can analyze financial statements, detect inconsistencies, and flag suspicious transactions that may indicate fraud or accounting errors. For example, AI can identify sudden, unexplained increases in spending, duplicate transactions, or discrepancies in vendor payments. Businesses can use AI to conduct real-time audits, allowing for proactive financial management.

Automated Invoice & Expense Management - AI can categorize transactions, track recurring expenses, and highlight areas where costs can be reduced. Many AI tools can auto-categorize expenses, flag duplicate invoices, and suggest budget adjustments based on spending patterns, improving efficiency and reducing errors in financial management.

HR & Recruiting

Resume Screening - AI can filter job applications based on predefined criteria, helping recruiters narrow down large applicant pools. However, AI-driven hiring tools have been found to perpetuate biases, particularly when trained on imbalanced datasets. Companies should regularly test

AI-generated hiring results against real-world outcomes to ensure fairness and mitigate discrimination.

AI for Employee Training & Onboarding - AI can generate interactive training modules and onboarding guides tailored to company policies. While this can streamline employee integration, human oversight is necessary to ensure that training content remains engaging and contextually relevant.

Healthcare & Wellness

Medical Documentation & Transcription - AI can transcribe doctor-patient interactions, converting them into structured notes that assist with record-keeping. This reduces administrative burdens on healthcare professionals, allowing them to focus more on patient care.

AI-Powered Health Insights - AI can analyze symptoms, lifestyle habits, and medical history to provide personalized health recommendations. However, AI misdiagnoses often occur because models are trained on datasets that underrepresent minority populations, leading to inaccuracies. AI-generated insights should always be reviewed by qualified medical professionals to prevent errors in patient care.

E-Commerce & Customer Engagement

Customer Service Chatbots - AI-powered chatbots can, and already are, handling customer inquiries, providing instant responses, and assisting with troubleshooting. By automating routine support tasks, businesses can improve customer satisfaction and reduce response times.

AI-Powered Product Recommendations - AI can analyze customer behavior and suggest relevant products based on past interactions. This helps businesses personalize the shopping experience, increasing engagement and conversion rates.

Real Estate & Property Management

AI for Lease & Contract Review - AI can highlight crucial clauses, potential risks, and discrepancies in lease agreements. This allows real estate professionals to conduct more thorough reviews without manually combing through complex legal documents.

Automated Property Valuation - AI can assess market trends, property characteristics, and historical sales data to estimate home values. These systems evaluate comparable sales, neighborhood trends, and economic factors to generate pricing estimates, helping real estate professionals make data-driven decisions on property valuation.

Key Takeaway: AI Can Be Your Custom Assistant

These examples showcase just a fraction of what is possible when AI is tailored to specific needs. This is where the beauty of creativity comes in—AI is not a one-size-fits-all solution, but rather a powerful tool that can be shaped and refined to fit your unique workflows, challenges, and opportunities. The more imaginative and strategic you are in leveraging AI, and emerging tech in general, the greater the impact it can have in transforming your work and industry.

AI's ability to streamline specialized tasks makes it an invaluable tool across industries, but its **limitations must be recognized**. While AI enhances efficiency, human oversight remains essential in fields where bias, ethical concerns, and contextual understanding play a critical role. The most effective AI users will be those who understand how to customize AI for their needs while maintaining accountability and ethical responsibility.

Practical AI Use Cases for Different Needs

Now that we understand how to structure prompts, and have a better idea of how some of these tools could be integrated, let's look at how this applies to different areas of work and creativity.

AI for Business & Productivity

- **Market Research:** "Analyze the latest trends in the AI industry and summarize key opportunities for new startups."
- **Business Plans:** "Generate a business model canvas for a subscription-based fitness app."
- **Automation Assistance:** "Suggest AI tools that can streamline customer service for a small business."

AI for Content Creation & Writing

- **Blog & Article Writing:** "Write a 1,500-word blog post on AI in healthcare, structured with an introduction, key trends, benefits, and future outlook."
- **Social Media:** "Generate fifteen Twitter (or X) posts on AI in entrepreneurship, each with a conversational and engaging tone."
- **Scriptwriting:** "Write a YouTube script explaining AI's impact on the music industry in a casual and informative way."

AI for Learning & Personal Growth

- **Skill Development:** "Create a self-paced 45-day learning plan to master AI fundamentals."
- **Language Learning:** "Translate this paragraph into Spanish and provide a pronunciation guide."
- **Career Growth:** "Suggest the best AI-related certifications for someone transitioning into data science."

These use cases show how refined prompts lead to better, more actionable responses.

The Future of AI Communication

As AI continues to evolve, so will the ways we interact with it. Mastering AI communication now will give you a significant advantage and understanding as AI tools become more integrated into daily life, which is happening extremely fast.

AI will eventually become more context-aware, meaning prompts will require less instruction over time. However, understanding how to ask the right questions will always be essential for getting the best results.

The Bottom Line: AI is a Skill—Learn to Use it Right

This form of tech isn't just a tool, it's becoming a completely new skill, and like any skill, it takes practice to use it effectively. Knowing how to write the right prompts, refine responses, and structure AI interactions is what separates casual users from those who truly leverage AI's full power. The future belongs to those who know how to use AI efficiently and strategically.

Understanding how to communicate with AI is only half the battle. Now, let's make sure you have the right tools at your disposal. In the next chapter, you'll find a curated selection of the best AI platforms for productivity, business, content creation, and automation—so you can start putting AI to work immediately.

09

AI Toolkit—Your Guide to Getting Started

Why This Toolkit Matters

AI is powerful, but only if you know how to use it. This chapter serves as a practical guide, providing carefully curated AI tools, best practices, and strategies to help you start applying AI today.

Whether you are a beginner looking for an entry point, a business owner wanting to automate tasks, or an advanced user exploring AI for growth, this toolkit will help you move from theory to execution.

Must-Have AI Tools by Category

Productivity & Automation

- **ChatGPT** - AI-powered text generation and brainstorming.
- **Notion AI** - Automates note-taking and task management.
- **Zapier** - Connects apps to automate workflows.

- **Grammarly** - Enhances writing clarity and grammar.

Business & Marketing

- **Jasper AI** - AI-driven content writing for blogs and ads.
- **Copy.ai** - Generates social media captions, emails, and ad copy.
- **Surfer SEO** - Helps optimize content for search engines.
- **Canva AI** - AI-powered graphic design for branding.

Learning & Research

- **Khan Academy AI Tutor** - Personalized education assistant.
- **Elicit AI** - AI-driven research tool for finding academic papers.
- **ChatGPT / Google Bard** - AI chatbots for general research.
- **Duolingo AI** - AI-enhanced language learning.

Finance & Investing

- **Wealthfront** - AI-powered investment management.
- **Mint AI** - Personal finance tracking and budgeting.
- **AI-Powered Fraud Detection** - Monitors financial transactions for security.

Creativity & Content

- **DALL·E / Midjourney** - AI-generated images based on text prompts.
- **Eleven Labs** - AI voice generation for narration and content.
- **Runway ML** - AI-powered video editing.
- **Synthesia** - AI-generated avatars for presentations.

How to Use AI Effectively

Having the right tools is one thing–using them well is another. Here's how to get the most out of AI without wasting time or feeling overwhelmed.

Start With a Clear Goal

Before using AI, define the specific task you need help with. Ask yourself:
- What problem am I trying to solve?
- Which AI tool best fits my need?
- How will this improve my efficiency or workflow?

AI works best when applied with intention. Randomly testing tools without a clear goal can lead to wasted time instead of increased productivity.

Learn the Art of Effective AI Prompts

Again, the way you communicate with AI affects the quality of the response you receive. A vague prompt often results in unclear or incomplete answers, while a well-structured prompt ensures AI provides meaningful insights.

Instead of asking, *"Help me with my business,"* refine your request to something more specific, such as:

"Outline a simple 30-day content strategy for my fitness brand on Instagram, focusing on organic growth."

Providing context, specifying a format, and clearly defining expectations will improve the usefulness of AI-generated responses.

Automate Repetitive Tasks

AI can help eliminate tedious work, allowing you to focus on tasks that require creativity and strategic thinking. Consider automating:
- Customer inquiries using AI chatbots.
- Scheduling and calendar management with AI-powered assistants.
- Summarizing lengthy emails, articles, or reports with AI text processing tools.

The goal is not to replace human effort but to enhance productivity by **reducing time spent on repetitive work.**

Use AI to Enhance, Not Replace, Human Judgment

AI is a tool, not a substitute for critical thinking. It is essential to fact-check AI-generated content, especially when dealing with:
- Research and news, as AI can occasionally produce inaccurate or misleading information.
- Financial decisions, as AI models cannot predict market trends with complete certainty.
- Creative work, where human originality and oversight remain irreplaceable.

Think of AI as an assistant that helps streamline processes while maintaining human oversight and decision-making.

Stay Updated—AI Changes Fast

AI technology evolves rapidly. New tools and features emerge regularly, making it important to stay informed. Keep up with trends by:
- Following AI newsletters, industry reports, and thought leaders.
- Experimenting with new AI tools and updates.
- Remaining adaptable to emerging innovations.

I can't stress this enough, the most successful AI users are those who continuously explore new capabilities and integrate them into their workflow.

Avoiding AI Pitfalls

Though AI is a such powerful tool, it is not without risks. Here are some common pitfalls to avoid:

- **Over-Reliance on AI** - AI should assist decision-making, not replace human judgment.
- **Misinformation & AI Hallucinations** - AI may generate incorrect information, requiring careful fact-checking.
- **Privacy & Data Security** - AI tools collect data; be mindful of sharing sensitive information.
- **Unethical AI Use** - AI should be used responsibly, particularly in hiring, finance, and decision-making processes.

By understanding these risks, you can use AI more effectively and responsibly.

How to Apply AI Based on Your Skill Level

If You're a Beginner:

- Start with ChatGPT for basic tasks such as summarizing content or brainstorming ideas.
- Use Grammarly to enhance writing skills.
- Try Canva AI for simple graphic design needs.

If You're a Business Owner or Professional:

- Automate email and marketing content with Copy.ai.

- Use Zapier to connect different business tools and streamline workflows.
- Leverage AI-powered market research tools for data insights.

If You're an Advanced AI User:

- Train custom AI models for specialized business needs.
- Experiment with AI-powered analytics and automation for scaling operations.
- Utilize AI-generated synthetic media for brand storytelling and content creation.

Regardless of experience level, there is an AI tool or strategy that can help improve efficiency and effectiveness.

Final Thoughts: AI is a Tool—Now Use It Right

AI tools provide the opportunity to develop a skillset that can be applied in meaningful ways to enhance productivity, creativity, and efficiency. The key to success is not simply knowing about AI but using it strategically.

A practical first step is to choose one AI tool from this chapter and begin integrating it into your workflow. Small, consistent applications of AI will lead to significant improvements over time.

The future of AI is not something to wait for, because it's already in your face and at your fingertips. Now, it's on you to lean in and take advantage.

Conclusion

The way we work, create, and build is evolving at an unprecedented pace. AI is way more than just another technological advancement, it is truly a fundamental shift in how we operate, think, and navigate through the world. Whether you are choosing to embrace it or ignore it will determine if you're staying ahead or getting left behind.

But let's be clear; AI isn't about replacing people, rather giving people the power to reclaim time and increase opportunity and impact. And if there's one thing we all know, time is the one resource you can never buy back. The difference between those who maximize AI and those who don't won't just be about skill–it will be about who is working smarter, not harder.

Imagine a world where the tasks that used to take hours can be done in minutes, well it's here. When research, content creation, problem-solving, and business growth can be streamlined with the right AI tools, what does that give you? More time to innovate, strategize, and lead.

AI is already shaping economies, industries, and the very structure of global power. The question isn't whether AI will be part of our lives–it already is. The question is who will control it, who will benefit from it, and who will be left out of the conversation.

That's why this isn't only about using AI, but it's about being part of the future AI is creating. Those who adapt will be the ones making decisions, setting trends, and creating wealth in the AI-driven era.

This book wasn't written to overwhelm you with theory. It was written to give you actionable tools and knowledge to start using AI right now. You don't need to be an expert, you just need to start applying what you've learned.

So here's your challenge: Pick one AI tool, one strategy, or one concept from this book and apply it today. Small, consistent actions will put you ahead of 90% of people who are still waiting to "figure out AI later."

And again, the future of AI isn't something that's happening to us, and isn't just about data and algorithms. It's about culture, creativity, and opportunity. And the best time to start using it? Right now.

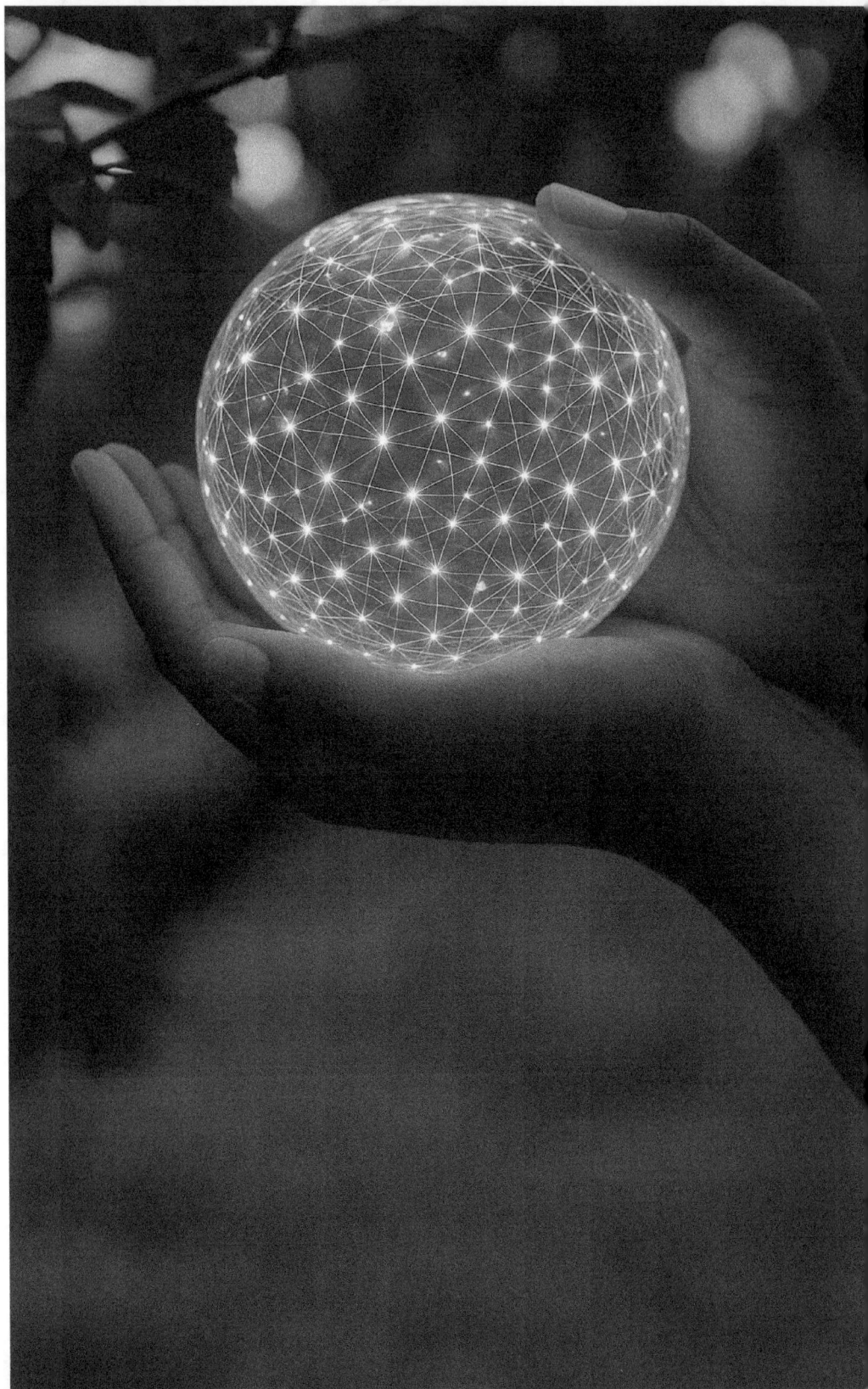

Key AI Concepts (Quick Reference Guide)

This section provides a brief, easy-to-understand guide to some of the most important AI terms. Whether you're new to AI or just need a refresher, these definitions will help you navigate the content with confidence.

Key Terms

Artificial General Intelligence (AGI)

A theoretical form of AI capable of understanding, learning, and performing any intellectual task that a human can. Unlike narrow AI, which is designed for specific tasks, AGI would possess general cognitive abilities, allowing it to adapt to new challenges without additional programming.

Artificial Intelligence (AI)

AI refers to computer systems that can perform tasks that typically require human intelligence, such as learning, decision-making, and problem-solving. AI powers tools like voice assistants, chatbots, and recommendation systems.

AI Bias

Bias that occurs when AI systems make unfair or inaccurate predictions due to flawed training data. AI bias can lead to discrimination in hiring, lending, and facial recognition.

AI Ethics
The study of moral and societal implications of AI, including privacy concerns, bias, and the impact of automation on jobs.

AI Hallucination
When AI generates incorrect or misleading information while sounding confident. This happens when AI fills in gaps in data with false details.

Automation
The use of AI to perform repetitive tasks without human intervention. Automation is used in customer service (chatbots), data entry, and scheduling.

Computer Vision
AI that enables computers to "see" and interpret images and videos, used in applications like facial recognition, medical imaging, and self-driving cars.

Deep Learning
An advanced form of machine learning that uses neural networks to analyze massive amounts of data. This is the technology behind AI-generated art, voice recognition, and natural language processing.

Generative AI
A type of AI that can create new content, such as text, images, music, and videos. Examples include ChatGPT (text generation) and DALL·E (image creation).

Large Language Models (LLMs)

AI systems trained on vast amounts of text data to understand and generate human language. ChatGPT is an example of an LLM.

Machine Learning (ML)
A subset of AI where computers learn from data instead of being explicitly programmed. ML algorithms improve over time based on patterns and experience.

Natural Language Processing (NLP)
AI that allows computers to understand, interpret, and generate human language. Examples include chatbots, speech-to-text software, and translation apps.

Neural Networks
A type of AI model inspired by the human brain that processes complex data. Neural networks are the foundation of deep learning and are used in everything from facial recognition to self-driving cars.

Prompt Engineering
The skill of crafting clear and precise instructions to guide AI models like ChatGPT to produce useful responses.

About The Author

Nyjal J. Drayton is a visionary innovator, entrepreneur, and advocate for emerging technology in underrepresented communities. With a background spanning business, innovation, and AI-driven solutions, Nyjal has dedicated his career to **bridging the digital divide** and empowering individuals to harness technology for economic and career advancement.

A dynamic leader in the intersection of **artificial intelligence, business strategy, and digital transformation,** Nyjal has worked extensively to integrate AI-powered solutions into education, entrepreneurship, and corporate ecosystems. As a champion for **HBCU students and young professionals**, he has been instrumental in introducing emerging technologies to institutions and organizations, ensuring that **our communities are not just consumers of technology— but builders, innovators, and leaders in the space**.

Nyjal's work includes pioneering AI applications in business development, content creation, and automation, helping individuals and enterprises **work smarter, not harder**. Through his hands-on approach, he has guided students, entrepreneurs, and professionals in leveraging AI for career growth, integration, and creative expression.

Beyond technology, Nyjal is passionate about **economic empowerment, generational wealth, and the future of digital entrepreneurship**. His mission is to demystify AI and emerging technologies, ensuring that everyone— regardless of background—has the knowledge and tools to thrive in the **AI-driven future**.

AI for the Culture: How Emerging Tech Can Level the Playing Field is a movement, not just a book. It's a call to action for readers to embrace AI as a tool for growth, opportunity, and change.

www.ingramcontent.com/pod-product-compliance
Lightning Source LLC
La Vergne TN
LVHW051815080426
835513LV00017B/1958